THE HATED HEROES

Vol 1

<u>My Dad is your Dad. AND IT SUCKS!</u>

Elliott Ryan

Elliott Ryan

ISBN: 978-0-578-50939-6

2019 Thank You For Remembering, LLC

DEDICATION

In loving memory of
James (Bunny) Camon
& his Mother
Mrs. Vera Camon

Early in life, I was privileged to learn the true meaning of Family. I grew up in a small, rural south Georgia town called Ray City. A large percentage of the entire population were blood relatives. They're an extremely close group of people. We were the only family that didn't have a blood connection. This didn't stop us from forming bonds as strong as that of blood relatives. A bond that exists today. I could dedicate an entire book to our experiences throughout those early years which molded those bonds. I just might. I'm typing this with the biggest smile while laughing and thinking about the late 80s and early 90s. Those Ray City years! Mrs. Vera opened my eyes to pure love and the true meaning of family. She always treated us the same as she treated everyone. When we were bad, which was often, she got on to us without holding back. I've always loved her for that. The way Bonny protected us, as he protected everyone is just...(deep sigh) there are no words to explain it.

Those lessons would be instrumental in the molding of my open-mindedness. In adulthood, these childhood lessons made it easier for me in the creating of similar bonds all over the world. Most of my adult life has been spent on the other side of the world. I would not be here today if I hadn't had those childhood experiences teaching me the true meaning of Family. To Lisa, Reggie, Ronald, and Deacon James, thanks and I will always consider you my family.

CONTENTS

Chapter 1

The Worst Hurt:

As kids, there are moments we all experience where we daydream about our future adult selves. We picture ourselves as the perfect fathers, mothers, and grandparents. Becoming a parent was always a responsibility I've held in the highest regard. It was the ultimate gift from the Universe! I often fantasized this illusion of parenthood being this magical experience.

I envisioned parenthood as this life-sized social gathering amongst adults where we would come together and share funny parenting stories. We would be drinking wine while exchanging valuable information about staying one step ahead of the kids. This would all seem somewhat unnecessary, seeing how well behaved all kids will be. Yeah. I was in for a huge rude awakening. It isn't until later in life you find out that alongside being the ultimate gift, is the fact that your mental and emotional strengths will be tested like you could've never imagined. Let's not forget physical strength. Those arms do get tired! As a [1]Platoon Sergeant and Senior Non-Commissioned Officer in the US Army, this is the story of the parental hardships I endured and the moment that destroyed my family.

We were at Lake Travis having a beautiful time. It was an impromptu barbecue for members of our unit and strangers at the lake with more beer than their party could finish off alone. This worked out perfectly because we had more fish than we could finish ourselves. Lake Travis provided activities for all age groups. The kids and women both utilized their designated areas. We menfolk had our area that composed of masculine conversation and whatnot. This area attracted my oldest son

---

[1] Equivalent to mid-career supervisory position in Corporate America.

Will and his friends. They thought they were of age to hang with and participate in our conversations. I was quite the storyteller therefore, I had everyone laughing.

I was telling the story about the time we took Sergeant Brown to a "club" for his 19th birthday. Brown was one of my favorite humans. I used him quite often in my stories. The Irony is, I'm telling this story about how good a time it was celebrating Sergeant Brown's birthday, while having missed numerous birthdays and special life events of Wills'.

While telling this story, a few of the boys began to gather their things and make their way to their cars. It wasn't a huge deal. The protective parent gene just drew me to call for Will. "Will, where are you guys headed"? He ignored me while continuing the conversation with his friends. "Will", I called again. This time Will turned, and in an aggressive tone shouted, “A PARTY”. A few of my buddies and I made eye contact when Will said party. The only other party we knew of was the LGBTQ party.

There was a moment of awkward silence as tense energy entered the area. My wife and her friends were engaged in their own conversations but sensed the tension and began to make their way over. We could see each other from afar, but couldn’t hear each other’s conversation due to the distance. I replied in a subtle but stern tone, “what party WILL”? By this time Will was 10 yards away. He turned back towards my direction and yelled, “MOM F******G KNOWS sheesh”.

Will finally reached his breaking point. He began to go off on the worst rant imaginable. It was a typical teenage boy crying out for attention from his Dad. With all there was, unknowingly suppressed inside me, it was the worst thing imaginable. “I DON’T HAVE TO

TELL YOU SH**! YOU'RE NEVER HERE. MY DAD HAS A VAGINA. THAT'S NOT CALLING YOU A B*** BUT MY MOM IS YOUR KING".

Will has always been such an articulate and smart kid. At the age of 7, he read and comprehended the classic book, Moby Dick. He always knew what to say, especially when he wanted a response out of you. After Will screamed, "YOU'RE NEVER HERE", my mind went blank. Everything began to move in slow motion. Will continues towards me, index finger pointed while biting his lip. Rightfully enraged with drool coming from his mouth, he continues to unleash his rant. I look up again and see my wife and her friend frantically waving their arms. They were screaming but I couldn't make out what they were saying due to the distance.

I refocus on Will. He is now directly in my face. He whispers to me, "Moms about to tell you". This is the moment our lives were forever changed. To understand this better, let me explain the events that led up to this gathering.

It was a Saturday morning and we were finally back home in Germany. I decided to take Will on a fishing trip. I sat motionless in a daze being charged by the Sun's beaming rays. I stared off into the clear light blue skies. We were fresh off a long 15-month deployment to the Middle East. The past few weeks my emotions were all over the place. This was also the beginning of a new cycle of my life. I was set to retire in 18 months and my oldest son Will had recently graduated from High School. I was able to attend via Video Tele-Conference from Iraq. Our entire team had been given a break to spend some quality family time.

Throughout my career, many of my soldiers were looked to as my family. I've always had a father figure, big brother, mentor-type relationship with my soldiers. My soldiers ranged from teenagers to young parents and they all knew I was a phone call away. Sometimes I was with them when the Police were called on us. I was looking forward to spending time with Will. I hadn't spent as much time being a physical presence in his life as I should have.

Unknowingly of Will's resentment, I used this time to try and strengthen our bond. I thought it was working. This particular day we were fishing at this secluded creek a short distance from Lake Travis. It was a popular fishing spot because it wasn't too far from the main Lake. It was also far enough not to be bothered by anyone.

As Will and I spent more time together, I began to feel the pressure of having to climb yet another mountain. The completion of my military career should've been seen as an outstanding feat. The feeling of joy was canceled out with the feeling of resentment from someone whom I had (financially) provided everything he needed to succeed in Life. I also began to think about life after I retire and having to find a job.

All of these emotions were starting to weigh heavy on my mind as we both sat in silence with our fishing rods in the water. I began to feel the need to be around familiar happiness and sadly, it wasn't Will.

In the distance, we began to hear a commotion coming from the Lake which interrupted our peaceful moment. The noise did bring me back from my daydream only to glance over and see Will sweaty, shirtless, and on his phone. It was the middle of the day and we've been out there for hours. It was an excellent day of fishing. We caught more than enough. Seeing how we were never going to eat all of that fish alone, Will suggested we call it a day. I could tell he was ready to leave. We packed up our things and began our ride out of the woods. The creek was about 2.2 miles down this bumpy dirt road which leads to the main highway. While traveling back towards the main highway, the noise began to get louder. The smell of barbecue began to fill the air even more. We were still surrounded by woods but the Lake Travis entrance was a short distance before you get to the main road. You could see the Park area when driving to and from the City.

Will suggested we stop by and give the people at the Lake some of our fish. The wooded area began to thin out as we were approaching the entryway for the 12 acre clearing.

I slow to make the left turn into Lake Travis and get a welcoming surprise. There enjoying themselves were Sergeant Brown and some of the other soldiers. Also standing there were the Company Commander and her husband.

"[2]*WHAT THA WHAT"*, I thought as I drove into Lake Travis thinking I was intentionally left out of a company picnic. I turned to Will to express my disdain only to see him uninterested and on his phone.

Truth is, Will had been texting his friends, informing them to meet up at Lake Travis. There was no intention of going back to the house to clean fish with me. His goal the entire time was to link up with his friends at the Lake. This worked out well for the both of us. I walk up with my arms extended, communicating nonverbally, "where's my invite"? Before I could get a word out, the Commander chuckled and said, "I was thinking the same thing". She and her husband had been driving along heading back into town before spotting the soldiers out there enjoying themselves. All the soldiers remaining at the barracks decided to

[2] Saying for when you want answers but are too upset to verbally articulate your demand.

go to Lake Travis to enjoy themselves.

While talking with the Commander, we decided to give a few other individuals a call to come out and join the fun. Before you know it, everyone that didn't go out of town came to the Lake with their families. We turned the event into a huge gathering and spent the rest of the day there.

Chapter 2

Some will say it's photoshopped!

The period was the early 90s and it was a typical quiet Saturday morning in my hometown of Ray City Georgia. Ray City was a small country town in which everyone knows everyone. There wasn't any vicious crime as you could sleep with your windows open and doors unlocked.

The time was approximately 555 am and one of my childhood friends came by the house. He stopped by my bedroom window, pushed the curtain aside, and told me to come out front. One thing about my childhood was that from sunrise to sunset, we were outside. We didn't play video games. We would be in the woods, swimming in the creek, or doing something outdoorsy. We were huge sports fans. We would go to the [3]Church Yard where

---

[3] Open grassy field/parking lot area for church parking that we would sometimes use.

we played pickup games of football and baseball. This particular Saturday, we planned to get up early and fix the basketball court in my front yard before everyone came over. We would pride ourselves on making a sturdy basketball goal. It had to be sturdy enough to dunk and hang on it (get all fancy without it breaking).

We assembled everything we were going to need to fix it. The only thing lacking was another backboard because someone ripped it down the previous day. That someone being our older cousins/brothers in the neighborhood. They would periodically kidnap our games and tear down our basketball court. They did this so we would get upset and fight back. At first we would cry. The crying was more so out of knowing we couldn't beat them, rather than the physical aspect of pain. It did physically hurt too. The crying just made them beat us even more. They would hold us down and punch us, repeatedly, in the same spot. It was some type of ghetto torture technique. While beating us, they would be saying we weren't growing up to be punks.

☼ Quick alibi.

(*Back then that word was used to describe a soft and weak individual (male or female) that couldn't fight, rather than the insensitive, offensive, homophobic way in which it is used today.*)

That was their way of making us tough. They would beat us until we stopped crying and fought back. It took us a while, but we found the secret to making them stop kicking our butts. All we had to do was fight back. When we would try and fight back, they ran away laughing. Occasionally they still beat our butts. For the most part, they would run off laughing. To this day, I still love them for those lessons. No matter how big they are,

FIGHT BACK! Never have victim energy! Fight back and the pain will stop. We all grew to become family. They were looked to as our older brothers and we still communicate to this day. Some have passed on but their spirits live on in the man I grew to be. This book is dedicated to one of them and his Mother. My childhood neighbors.

Alright. Back to the story. We searched in our immediate area but couldn't find anything sturdy. We decided to jump on our bikes and ride around the neighborhood in search of some more sustainable plywood.

Our journey took us to the other side of town because our friend's Uncle was into construction. We figured to "borrow" some good wood from over there.

My friend and I would get sidetracked easily. As we were riding around town, we noticed our other friend's car. They were from the next town over so we got excited. We knew there would be more people over to the house playing basketball. We rode by the car and, at that very moment, The Friends Dad came out the house to throw some trash away. He was shirtless with these leopard print underwear and some fluffy pink sandals.

To know him was to understand this outfit. We always called him a pimp. He was a tall and lanky, hot head, alpha personality individual that took no shit. He spoke with a slow southern draw like he was always drunk. He was indeed always drunk.

We were young so we whispered some jokes amongst ourselves. He looked at us and asked what we were doing around there. We told him and he replied with a chuckle, "so you little niggaz about to steal LC wood". He then told us to come in. I looked at my friend, he looked at me and we were like, "Nah we got to get back across town", with this chuckle in our voice. He replied more sternly, "YOU LIL

SCRUBS COME THE F*** IN". He walked in, leaving the door open. He was family so we weren't creeped out or anything. It was more so that terribly good feeling. There could've been anything in there. We figured he was going to;

(1) Give us some money to buy a basketball goal.

(2) Show us an extraterrestrial being.

(3) Help him dispose of a dead body.

(4) Teach us how to cook dope.

You just didn't know because he was one of those characters.

➤➤➤Over the following years he showed us all those things.

You enter the house through this narrow

entryway. The kitchen is to your front-right. There's a door leading to a master bedroom on the far side of the kitchen. To our immediate right was a dining room table full of empty liquor bottles, cigarettes, guns, and a whole bunch of other drug paraphernalia we couldn't make out at the time.

The living room was to the direct left. There was a tall china cabinet between us, blocking the full view of the living room. We could see The Dad because he was sitting in this recliner chair to our left-front. He told us to come on around. We continue into the living room and out the way of the china cabinet. You could also see down the hallway to the right behind The Dads chair. This hallway led to a couple bedrooms, a bathroom, and a back door. You can start to see what looked like a leg in the living room.

As we walk around the Dad's chair, we would block his view of this floor model television placed against the wall. There was a picture of a Black Jesus hanging on the wall. We could see out this window into the front yard where our bikes were parked. On the other side of the china cabinet and under the window was a love seat sofa. As we focus our attention on the sofa, we get the shock of our lives. Laying right there on the couch was this woman in her birthday suit.

We both stood there staring at her in a complete state of shock. Speechless. Her head was on the arm of the sofa. Her left leg was on the back of the sofa, with her right leg dangling on the floor. It took about 30 seconds before I looked at her face and realized who she was. To know her was to feel like we were in Heaven. The first words out her mouth were, “So you boys never saw any vagina before”? The answer to that question should’ve been no. I replied, “Yes Mam”. The Dad then replied, “You little scrubs ain’t never had no vagina”.

As we were standing there, she told one of us to come here. I looked at my friend. He looked at me. We never said a word but the eye contact was, YOU GO. NO, you go. NO, you go. He then pushed me in her direction so I continued towards her. As I walked up to her, the feeling start to go from amazement to scary. It wasn't the scary feeling one would have from an enemy or a threat. It was the scary feeling of not knowing what to do or expect. She was about to find out I had no clue what to do! We've watched "movies" and had some "magazines" but never had we been with a woman before.

I reach her, she sits up and puts her legs on the coffee table, straddling me. By this time, my heart is racing and my stomach is in knots. I'm with a grin from ear to ear. She asked if I know how to eat a woman. Well, I'm extremely nervous so what I heard was, "*Do you know how to eat with a woman*"? I replied, "yes

Mam. You pull out her chair, let her sit first". I began explaining how to be a gentleman when on a date with a woman. She interrupted my explaining, replied with laughter, "Not eat with a woman silly. Put your face in some vagina". I turn to look at my friend and he's back there laughing. Of course I lied. I then looked at The Dad and he hasn't paid us any more attention since we got in. He was all into the TV show Good Times.

She then told me to get on my knees and show her how to eat a woman. I'm extremely excited but was very reluctant to do it because obviously, she was about to find out I didn't know what I was doing. I got on my knees and the first thing comes to mind was a movie we watched. I recalled the man head moving in a circle as he was kissing and licking. I figured as long as I did that, I would be good. Moving forward, the only thing going through my head was Lick ➤Kiss ➤Circle. Lick ➤Kiss ➤Circle

➤Lick ➤Kiss ➤Circle!

Getting closer, I realize there's a lot of moving pieces to this thang. I was expecting this simple hole, not all these meat particles everywhere. To top it all off, I didn't see a hole where I expected it to be. "Is this some kind of joke"? "Where is it"? My mind was racing. This was all beginning to be too much. Once almost there, I was terrified. I closed my eyes, puckered my lips and waited to feel flesh.
I stopped soon as I touched skin. I felt the right side of my face blocked so I only opened my left eye. I realized I wasn't even on it. I was on her inner left thigh. With only my left eye open, I get the shock of my life. It was like looking at a vagina in a microscope. All I saw was what looked like juice. It crossed my mind that she tricked me into licking her pee. She let out a chuckle then gently placed her hands on both sides of my head to reposition it center mass on her vagina! It didn't smell like pee.

There was a watermelon smell. This made me think more so that licking was the appropriate thing to do.

At this point, my mind goes back to the movie I watched and I say Y.O.L.O. (Well not really Y.O.L.O. because that phrase wasn't invented then but you get the point.)

With my eyes closed, I began rotating my head in a circle while kissing and licking. I also remembered the sounds they would make in the movies and figured if I start moaning that would make it better. So I began to moan while licking and kissing with my head going in a circle.

I didn't think I was loud at the time but later on my friend told me I was so loud people would've been able to hear me outside. After what seemed like an eternity, which was more like 12 seconds, she placed her hands on both sides of my head to lift it up. She says to me, "Honey, what was that"? I was so happy and excited and what I heard was, “Honey that was awesome”. Grinning from ear to ear, I turned and looked at my friend. We mentally hi-five each other while nodding our head in total pleasure. As I’m looking at My Friend, he turns and makes eye contact with The Dad. His facial expression immediately goes from happy to scared as if he’d seen a ghost. I turn and made eye contact with The Dad and immediately removed the grin from my face also. I make a deep swallow and thought to myself, oh shit.

The Dad had turned the TV off. There was this blank look on his face. He then shakes his head in displeasure and says, “Lil nigga that’s got to be the dumbest shit I’ve ever seen. WTF are you moaning for”. She then stands up, points at the sofa and tells me to lay down. I’m thinking, well I’m not thinking. My mind is blown and I’m reacting to commands. I get on the sofa, lay on my left side with my back facing her. I position my hands together and placed them under my head as a pillow. She looks down at me and says "what are you doing"? "Laying down", I replied. She then says, "ON YOUR BACK SILLY". I look at The Dad and he's looking at me with the same displeased look and say, "Elliott! You're that special stupid"!

I turn on my back. She kneels beside me and grabs my face. She then begins to explain the proper procedure.

Here I go being dumb again.

As she gets closer to my face,

I close my eyes.

I halfway open my mouth.

I pucker my lips and stick out my tongue like we're about to have this sloppy kiss.

➤➤Take a quick second to literally make this facial gesture and see how stupid I must have looked.

She stops talking so I open my eyes with my mouth open and tongue still out (looking dumb). She was looking at me with this smirk and says, “what the f*** are you doing honey”? Before I could say anything, she quickly replies, “close your mouth and listen”. I look at The Dad and he had the same displeased look on his face while shaking his head in disappointment! She proceeds to explain the female vagina to us. While explaining the layout of the female vagina, she’s opening and closing my mouth with her tongue. She begins explaining how the vagina has lips just like your mouth and so on and so on.

Chapter 3

Be careful what you ask for

Some would say that genuinely good humans are rare. I tend to believe this world is full of genuinely good humans. Those individuals don't get acknowledged as much as they should. A perk of being in the military is crossing paths with countless humans from all walks of life. You will grow to love and form unbreakable bonds with many genuinely good souls. Timothy and Lucas were two friends who met at a Military In Processing facility. This is the place where you go and get your physical screening once all the initial paperwork has been finalized.

This is essentially the place where many bonds are formed. Some guys, as in the case of Timothy and Lucas, find out they will be attending Basic Training at the same location. Timothy was a little older than Lucas. He didn't join the Military immediately after High School. Timothy was more reserved and didn't care about being the most popular as life hit him like a ton of bricks at an early age.

Lucas was an energized high school graduate. He was anxious to get away from home and experience life more. A fiery Sagittarius Sun sign. He was very outgoing and full of jokes. People either liked or hated him. No in-between. Lucas was an extrovert to the very definition. A real "social butterfly" as he was called on numerous occasions. Lucas was the only child to an elementary school teacher and a police officer.

Their home was the typical small-town suburban community with little to no crime. The city police officers knew all the neighborhood kids' names. Lucas and his friends would often play those violent war video games and the street gang video games. This would add some excitement to their seemingly privileged lives. During free time, the guys would crowd around in the dayroom and play video games. Lucas, being the video game expert, would be the loudest person in there. He was always going back and forth with semi-friendly banter.

Most of the guys liked Lucas. There were indeed some that COULD NOT stand him. Lucas really took to Timothy. Quite often when in heated moments playing video games, he would call on Timothy so he could get that big brother synergy.

As their time at In-processing was winding down, Timothy and Lucas received good news. They would be attending Basic Training together. Their friendship also became stronger (more so in Lucas' eyes) when they found out they both decided to attend Army Airborne School. Lucas' decision was made out of being a super hyper-energetic young kid with a taste for travel and war. [4]Tim's decision was for financial gain. They offered him bonus money and he said ok. After In-processing, they exchanged numbers so they could stay in touch before leaving for training in the coming weeks.

Basic Training and Army Airborne Training (Jump School) was everything Lucas thought it would be. He loved it. Often during training, Lucas ended up on the wrong side of some pretty intense physical skirmishes. This neither deterred nor frightened us (of course I helped my friend fight). Timothy admired the fiery side of his friend. Lucas was always unapologetically himself. He was of supreme confidence. He operated on a different frequency than most. He thought, in his heart of hearts, that those that didn't like him didn't listen to him enough and needed to spend

---

[4] Abbreviation for Timothy as this is how I will reference him for the majority of this book.

more time around him. Lucas was this ball of young, eager, enthusiastic positive energy. He thrived with the disciplinary and the firearms training. He viewed this area of his life as needing more improvement if he wanted to embody the ultimate soldier he envisioned himself becoming. It was during these moments in training where Lucas would realize this was where the void in his life lies.

Training had the opposite effect on Timothy. It validated his thoughts of the Military prior to signing up. Tim knew he could handle the training. He'd already made up his mind. He figured that unless things would be tremendously different at his duty station, he was getting out after his tour was complete. Tim came to the understanding that the closure he was looking for wasn't coming from his time spent in the Army. Despite their opposite views of the Military, their friendship continued to grow. They would quite often acknowledge each other as being very instrumental in the success obtained in their early careers. Their bond grew into that of brothers.

After training, Timothy spent most of his time at Lucas's home. Tim relocated after high school to escape some past demons. Lucas's family was more balanced and loving. There were also more events planned at Lucas' home. Lucas's Dad grew fond of Timothy. He made sure Lucas would always bring him over. Lucas didn't understand why his Dad asked about Timothy so much. He basically took Timothy in as a son. I guess it was his alertness of being a Police Officer which led to his sensing of Tim's energy of longing for love and compassion. Nevertheless, this was further assurance for Lucas that Tim was indeed his brother and he loved him as such.

Lucas would be bragging about the training and how easy it came to him. Blah blah blah thought Tim. He would often get tired of hearing Lucas gloat. Tim would counter Lucas's hero claims with this hilarious story of Lucas being, not so manly, during Airborne School.

This event almost ruined their friendship. This would have caused their careers to take totally different paths. Tim enjoyed this part of their friendship! He had the only, THE ONLY tool known to man that could stop Lucas mid-sentence and have him speechless. Tim knew he could always get his buddy to hush on command. Lucas would always in the most pissed little brother tone ask Tim, "Why you always bringing up old shit". Lucas hated that story. It would highlight a hilarious flaw in his manliness. He knew he would have to fix this to become the "ultimate soldier" he envisioned himself becoming. Yeah, whatever. He was just embarrassed.

This story was Tims ULTIMATE bargaining

chip whenever there was a rift between the two and they couldn't come to an agreement. Needless to say, Lucas would lose and Tim always got his way.

Tim and Lucas had different jobs but were extremely lucky. They were both assigned to the same Company. As with the nature of their friendship, where there's good news, the bad news is sure to follow.

It was also during this time when the Nation declared war and their company was tasked with being one of the first to deploy to combat. Lucas became even more excited. His whole being was to get more proficient in firearms.

He envisioned himself running, gunning, dodging bullets all while rescuing innocent puppies out of the grasp of "terrorists" like in the movies. Much to his disappointment, the first deployment came to an end without Lucas having fired his weapon in actual combat. To add further insult to Lucas's disappointment, our company loss 9 soldiers during that deployment. The Army became his everything and their deaths were personal. Lucas hated the fact that his job wasn't a combat job. He hated even more not knowing this information prior to entering the Army. Lucas vowed to change his job to one of the Army's Combat jobs.

Tim found some excitement in his job. It involved monitors, radios, drones, helicopters, and many moving pieces in Aviation. Tim didn't mind not having a Combat job. His early life experiences exposed him first hand to the effects of gun violence and the unpleasant ruggedness of life.

Upon return from their deployment, it would be time for Tim to begin his separation process as his 4-year contract was coming to an end. As promised, Lucas filed paperwork to change to one of the Army's Combat jobs.

Their careers and lives were beginning to move in opposite directions. This never hindered their friendship. Lucas would always seek guidance from Tim. Tim would always check in on Lucas to see how he was doing. One of the tell-tell signs for Tim was that Lucas would be playing video games while telling jokes and being the life of the party! This always let Tim know his friend was doing ok.

Years passed and the communication between the two friends had gotten scarce. Their feelings towards each other never changed. With the war having escalated, Lucas would deploy back to back. During those times, it was normal for a Combat soldier to have spent 3.5, sometimes 4 out of 5 years in a combat zone. Lucas excelled up the ranks, having become a top soldier. He gave the Army his all in becoming the Ultimate Leader. He developed a mission-first attitude that was perfect for his team and his soldiers. Lucas always talked about his friend. He would often use stories of times with Tim when explaining to his soldiers the negative outcomes and consequences of doing stupid things. There were photos of himself and his brother on the desk in his office. To say the least, everyone who came across Lucas's path knew of the importance of his brother.

Tim bounced around from job to job and never really grasped his life's purpose. His highly sought after closure would seemingly forever evade him. Tim's military experience made him resent the Army more. He missed his friend and viewed the Army as taking yet another necessity from his life. The two friends would eventually lose touch. The War left Lucas with very little time for family and friends. Lucas would also move on to serve in some of the more Elite facets of the Military. He gave the Military his everything.

Years passed and Tim finally pulled himself

together. He would get a job that required his travel to the Middle East. His job required him to supervise an aircraft crew as they would fly around Afghanistan. This particular day, Tim and his crew would be staying overnight at this FOB in Jalalabad Afghanistan. Tim was not pleased. He was tired of flying and wanted to make it back to Kabul and sleep in his own bed. He knew the sleeping quarters would be full of noisy soldiers.

Tim and the helicopter crew entered their temporary housing unit. It was as he thought it would be! Soldiers enjoying their free time while playing video games, shooting pool, and being loud obnoxious soldiers. Tim never forgot his friend and would always look for him as he traveled throughout the Middle East.

This particular time, Tim went to see the source of the madness. As if the Universe was going to place Lucas in that dayroom. Lucas was not there being loud playing video games.

Tim left to put his bag in his room. After a few hours, Tim realized he wasn't getting any sleep that night. Tim made his way back to the dayroom to hang out. The living quarters were this approximately 2400 square foot rectangle shaped wooden hut. There were 7 rooms on one side and 5 on the other. Some of the rooms had multiple bunk beds that held many soldiers. There was a long hallway that stretched from the back door to the front of the hut. The day room was this open area to the front right of the hut.

If you entered through the front entrance, you would have to walk through the madness to get to the back rooms. Tim approached the dayroom from the back entrance as he wanted no parts of the video games and horseplay. The back area was separated by this wooden white picket fence looking ordeal. This provided a little privacy to those sitting back there in the recliners and sofas. The lighting was dim and suddenly this blown bulb started flickering. This empty area caught Tim's attention. The set up was like that of a horseshoe. There was a 3 seat black leather sofa with empty recliners on both sides. On the sofa were two soldiers sitting together. The older soldier sat with his right arm around the younger soldier.

As Tim approached from their left, the younger soldier looks up and Tim could see the soldier had been crying and was distraught. The young soldiers' eyes stretched wider as he made eye contact with Tim. It was as if he knew him. The older soldier then looked over his shoulder to see who caught the young soldiers' attention.

Sitting right there was Lucas. He was consoling one of his young soldiers. One probably came to a premature conclusion that this reunion was a fairytale where Lucas jumps up, gives his good brother a humongous man-hug, kick back and do a lot of catching up. That was not the case. Tim immediately knew something was wrong. As the two friends made eye contact, Lucas merely extends his left arm, reaches for his friend's hand and sat him down beside him while continuing to talk to the distraught soldier! Tim could hear Lucas finishing off positive affirmations the two friends recited together numerous times after dealing with hardships. Tim also finished off the positive affirmations under his breath.

After a prayer, the soldier wiped his face and says to Lucas, "thanks Sir". They stand up and, before the young soldier would leave, an older gentleman came over. Lucas shook his hand and said to the older gentleman, "Sir he's going to be alright", referencing the young distraught soldier. The older gentleman was the Commander of the [5]FOB. The Commander made eye contact with Tim. He looked away

---

5 Forward Operation Base

then did a subtle double-take. He then stared at Tim for a moment before leaving. Tim thought this was again extremely weird but almost immediately forgot about it as the two friends would then give each other that fairytale hug. Lucas had his own more private sleeping area where he and Tim had gone. Lucas got his friend a chair, sat down and almost immediately begin to sob. Lucas began telling Tim about the troubling deaths of many of his soldiers and innocent Iraqis whom deaths were considered "collateral damage". Some of the more gruesome stories one could imagine. Lucas and that particular soldier were the only two from their Unit there at the FOB. They were accompanying the body of a deceased soldier to Mortuary Affairs there on Jalalabad. Lucas talked and cried and cried and cried all night long. He described how he had gotten what he wanted, only to realize it was something he couldn't handle. Lucas had years of suppressed hurt as he always felt the need to keep up the funny out-going persona. He was the person everyone came to when they were hurting. There was no one to console him. The morning came and Tim was very reluctant to leave. He could see what everyone else couldn't. The overwhelming pain in his brothers eyes! Tim felt there was something else his friend wanted to tell him. His intuition sensed this the whole night and he was waiting for Lucas to bring forth the secret. Lucas never did tell him. Tim hugged his friend and they parted ways as he now had to rush to get the helicopter crew back to Kabul before the bad weather moved in. Tim left not knowing this would be the last time he would see his friend alive. A few days passed before Tim could get back to Jalalabad for what he thought would be a reunion with his friend. Tim was unfortunately mistaken. Lucas had taken his own life. Tim made his way over to the memorial site the Base set up for Lucas. The site was serene. From its high elevation, it overlooked the valley below as you stared into the cloudy mountain tops. While standing there, this older gentleman came over. Tim glanced and seen it was the Commander that came and talked to Lucas the night he was consoling his soldier.

After a few seconds of standing there in complete silence, the Commander turns to Tim and says, "you don't remember me do you"? Tim looks at the Commander's name tag and rank and then replies, "No Colonel Brown, I don't". The Commander replies, "You can call me Brown, I'm Sergeant Brown from Lake Travis, your Dads soldier". Tim takes a step back, looks in the Commander's eyes and says, "you can't be". The Commander then says to Tim, "I'm sorry about your parents Will". Your father is the best human I've come across in this lifetime. Your father forced me to put together my Officers packet. He thought I would make a great Commander someday. Tim replied, “why didn’t you call me out the night with Lucas”? The Commander replied, “I’m going to let you continue to hide if you want”. Will asked, “Do you think the death of Mother that day, and Father that night, is my fault? Do you think about that day often”? The Commander replied, “I’ve tried, to no avail, to erase that day from my memory. The piercing of the air from your mother’s final scream! The look in Dads, I mean, your dads’ eyes, wakes me up every night.

Whenever I do find myself replaying that day, it always begins where you whispered something to your father. I was making my way towards you two when you suddenly punched him to the ground. From my vantage point, I could see your parents bodies hitting the ground at the same time. It’s like I was seeing it all happen in slow motion. Your

Dad's head began rotating towards your Mothers scream while he was falling to the ground. The punch to the face and your Mothers scream happened simultaneously. I made it to your dad around the same time your friends made it to you. The piercing scream of your mother's last cry dominated all other sounds in the park that day. Everyone in the park stopped what they were doing, only to focus on your mother as she had collapsed on the ground. It has always baffled me what it was your mom was trying to say".

Will replied, "I legally changed my first and last name. She wanted to tell Dad first. I think she sensed I was about to break the news to him. The autopsy shown that her heart burst from the stress of the moment. She was dead before her body hit the ground". Will apologized to The Commander for being jealous of the relationship between him and his father. Will explained how his spiritual journey has helped him to understand the power of pure love. He explained how he understands how his Father can love others the way he loved him. **Love is a Superpower**. The Commander then asked if he could take a picture of Will there at the memorial site. Will had tears in his eyes forcing him to look off into cloudy mountaintops as the Commander snapped the photo. Afterwards, the Commander says to Tim, "You know, Lucas's dad was a good friend of mines. Your dad saved his life. Think about that for a moment". The Commander reached into his pocket and gave Will this handwritten note Lucas wrote for him. The Commander then turned to leave the area, leaving Tim alone with the letter. Lucas' note read as follows;

My brother, I have a lot to share with you but let me start by telling you that my Dad has known your true identity since the first time you came to the house. He and my mom served in the Army under your father for a couple of years before getting out and joining the police force. Just know………

Elliott Ryan

The Dilemma

!

Tim had been sitting on the side of his bed praying. The pain weighing him down like a ton of bricks suddenly intensified as he began to process his friend's journey into the afterlife!

Tim laid back on his bed, put his hands over his face, and began to weep harder. Tim thought about how in his childhood teachings of religion, suicide is a sin which you cannot ask for forgiveness. Therefore, Lucas's actions solidified his Eternity. As Tim began to think more about this, he began to get angry at his friend.

Tim thought about Lucas' final moments. He drifted off thinking about conversations they would have when their friend got killed in a car accident. They would talk about how they would be reunited someday in Heaven (or jokingly, in Hell). Lucas took his life understanding their religious teachings of how suicide would assure a ticket to an eternity in Hell. Tim viewed this as Lucas not only ruining his own life. Tim saw this as Lucas selfishly attempting to ruin his life also! Lucas knew he was all Tim had.

Tim regretted his brother not joining him on his spiritual journey. He pleaded to Spirit, our Divine Mother of Creation, for mercy on his brothers Soul.

P1 - Afghanistan overlooking the mountaintops and valley below.

Chapter 4

The Move to Texas

The time was around 1996 and it was the first day at my new High School. It was the start of my Junior year (11th grade) and I moved from Georgia to Texas to stay with my Dad. My parents divorced and my Dad moved to Texas. Early in life, I decided I wanted to see the World and envisioned this move as the beginning of doing just that. This was especially hard for me, seeing how I was described as a "Mommas Boy" by some of my younger flings.

Texas wasn't new to us as me and my brothers would make friends during previous trips there. We have family in Texas, so the move wasn't too much of a shock factor. This day in particular was bittersweet as it was the first time I was in Texas without my two older brothers. It was the first time I was about to face the madness all alone. They were both out of school and back in Georgia with one having joined the Marines and the other in and out of trouble with law enforcement. I was terrified, but not a cowardly scary. It was instilled in us at an early age to be the same "gangsta" you were alone as you were with your brothers. I use the word gangsta very loosely. We were raised by some gangstas but not raised to be gangstas! (*Not all the gangstas raising us were males.*) My brothers and I lived these teaching to our core. One of my favorite songs goes a little something like......[6]*Every time I see you cats/you rolling in packs/for the life of me I cannot see why you don't know how to act/ love to clown when you deep/but when you on that solo creep out on the street you don't hear a peep......* whoever finished that verse, you and myself would, more likely than not, enjoy a

[6] Lyrics from the song Homeboyz off the 2Pac & Outlawz 1999 album - Still I Rise

future joint together.

I'm back. I got carried away. Every time I think of that song, I have to sing that verse. Back to the story.

This was the second day of school. The previous day my Dad's car had malfunctioned, to say the least. My Dad had this 1986 money green Jaguar X type that would turn heads everywhere we went. It was an older model but it was a Jaguar non-the-least. The issue with the car this day was the dipstick for checking the oil was missing. I still don't know what exactly happened to that dipstick. Of course it was my fault. Not having a dipstick would let oil spew onto the engine while driving. This would cause the engine to be smoking like a chimney. We spent the previous day looking to replace the dipstick but couldn't find any other that could fit. The cost to replace it also factored into the decision of my Dad to say, "well ya know the engine isn't smoking that bad"! I was already registered so there was no need nor want for my Dad to take me to school. I roamed freely during those times. I pulled into the school parking lot and some looked in amazement at me driving a Jaguar. Others chuckled and hurled jokes as you couldn't help but smell and see the effects of motor oil being spewed onto a hot car engine. I was still with more of a blank feeling as I sat there in the car watching people go into the school. I sat there and began to daydream about the previous weekend when I was at my friend Joone's home for a back to school barbecue. Joone was a friend I made during that summer. He stayed on the Northside of the city but my relatives stayed on the South Side. The gang banging was very bad during those times as the South Side was associated with Red and the Northside wore blue. My favorite color has always been blue. All the clothes I had were blue. I owned no red. I always thought it was too loud. I also associated Red with a terrible color on a suit. Deacons in the church would sometimes wear this which I thought was hideous.

I’ve always had an introverted personality on the surface. I just enjoy listening. I look at it as a skill to be able to nonverbally get your point across. While at the BBQ, I was sitting alone in a chair eating. Joone and some of his Blue friends were near talking and laughing. After a few minutes of eating by myself, Joone's Uncle comes and pulls up a chair. Joone’s uncle was an old head Gangster and had been banging his whole life. We sit there and he talks to me for a few minutes, dropping some jewels. Finally noticing us two over there talking, Joone walks up. When Joone gets close enough to hear us, the Uncle stands up and says to Joone, “Nephew, the Lil homie here pretty cool. He don’t try too hard nor say too much. He ain't all friendly either”.

As I stand up, The Uncle daps me up and says, “be easy Nephew and watch your back at school”. When I snap back into it, I realize the bell had rung and I was the only one outside! I get out of the car and begin to walk up to school. This school was massive. You could fit my entire hometown into this High School. The school was 3 stories and about five thousand students.

Walking into the school, I was beginning to feel overwhelmed. With each step closer, the sensation of being alone without my brothers begins to hit me harder and harder. I reach the school and enter the hallway. Only a few students were wandering around as the bell had rung. I was about to be late on my first day. One of the first things I remember was how long that hallway was. Crazy as it may seem, there was a North Side and South Side to the school.

I entered through the North entrance but

my homeroom class was at the opposite end of the school. I begin to walk down the seemingly never-ending hallway. While walking I passed this gorgeous female wearing this red body tight dress. We made eye contact and she looks me up and down. She refocuses on my eyes and shakes her head in disappointment and began showing me some hand gestures. I looked down at my shirt to see if I spilled food on it. This was the first time I consciously evaluated what I was wearing. I realized I was pretty much dressing like I chose to be down with the blue gang. I was wearing my favorite shirt ever. It was a blue Chaps shirt with grey stripes and some blue jeans. I didn't care too much as I knew I wasn't a Gangster and thought that was all it took. Yeah. About that! By this time I was about 3/4 of the way down the hall before I start to hear commotion coming from the higher floors. I continue down the hallway and the commotion gets louder. I hear a loud bang and the door I entered had violently burst open. I turn around to see two cops running full speed in my direction. I

move to the side as the officers pass by and continue down the hallway. At the end of the hallway was another set of stairs. It had gotten pretty evident that there was a brawl and I was about to meet it head-on! My homeroom class was up those stairs on the second floor. I make it to the end of the hall and begin to go up the stairs. Everything begins to move in slow motion as I hear the authoritative tone of a cop saying to me, "GET OUT THE WAY. MOVE TO THE SIDE". The cop was coming down the stairs with one guy in handcuffs. This dude was pissed. He was wearing red pants and a white and red striped, partially ripped shirt. I stood there against the wall and he didn't even notice me. He was more focused on the officers and attempting to get back up the stairs. I stood there looking at the second set of officers bringing down another guy. As the cops begin bringing the second guy down the stairs, I see his red pants first. I then begin to see that he did not have on a shirt. He was covered in tattoos. The energy was as if we were sensing each other's presence. From the moment he got into view, our eyes locked. He was in handcuffs with a cop to his front and one to his back. I noticed blood pouring from a wound on the side of his head dripping onto the ground. As they begin to make their way down the last set of stairs, the officers slowed down to get everyone gathered.

On the ground floor, the cops stop the guy directly in front of me. Literally. It was as if I was invisible to the Cops. I remember saying to myself, "umm Mr. Policeman, you guys wanna get baby Suge Knight from in front of me before he head butts me to death".

The guy gets in front of me, looks me head to toe, makes eye contact again and says, "I hope you're ready for this shit Blood". It was at that very moment where, if I had any doubt before, it was assured that I was all alone in Texas. The school was Central High in Beaumont (BMT aka Big Money Texas) Texas.

Only to do this all over again my Senior year in High School when my Dad got another job and we moved to Hiram Clarke Texas (Houston Texas) where I would graduate from James Madison High.

Fast forward a few years and I joined the Army. I had been at my Fort Bragg, my first duty station, a couple of years before it was time to do a tour Overseas. I received orders to go to Germany. I was very reluctant about going because I really loved Fort Bragg and knew I was going to miss my friends. This is one of the downsides to being in the Military. I was also coming out of a broken relationship and was a little sad (insert hilariously crying laughing emoji).

This was the first week at my new duty station in Hanau Germany. I was still extremely upset about being there. I was without my personal belongings and my Truck hadn't made it in yet. I decided to go to the gym and watch the guys play some basketball. This would add more fuel to the fire. I wouldn't be able to play because my gear hadn't arrived. It was either go to the gym or sit in my room. I was tired of staring at the walls so I made my way to the gym.

The gym was super packed and the two teams were down on the far end of the court arguing a call. There were probably 30 dudes and females in the gym. They were all on the sides of the court. I stood out because I was this new guy standing all alone in the entranceway to the court. After the arguing was over, the teams began to make their way back down towards me. The point guard was bringing the ball up the court. He was dribbling along the side where I was. As the guy makes his way down across half court, we make eye contact and he looks away at first. He then does a double-take at me. He slows down and comes to a complete stop with his hand motion still dribbling but the ball had, unknowingly to him, bounced away.

Immediately both our minds went back to our past interactions! I was already pissed and did not want to be there! I was ready to set it off. We both knew exactly whom each other was. The guys were calling him but we were both in a zone, nonverbally communicating with each other. Everyone in the gym stopped and was like "Yo Sham what you doing man"? He walks up to me and says, "Kuntry. From Central". I say, "watup Sham". He daps me up, hugs me and says, "man it's good to have my fellow Big Money Texas homie here". I say to him, "man I see you've changed". He smirks and says, "I see you have too". He introduced me to his homies. We all went out later. We chopped it up about old times. We would become good friends. Nobody knew of his past as he wanted me to keep it amongst ourselves. He didn't want anyone to know the level of Gangsta he was. Yeah, he was a somebody in the gang world. We both laughed about those times. We reminisced about those that passed and just thanked the Divine Spirit that we both made it out alive. The gang banging was very extreme back then. I told him about the one story that came to mind.

It was the middle of the day and my Geometry teacher would let me out of class early. She didn't care because Geometry was second nature to me. Math is my favorite subject. I was her class helper. She gave me an A because I knew the work. Geometry was super easy. The class was on the second floor. I began to make my way down to the first floor. While walking down the stairs, some spit landed beside me. I stop look up and see it came from Sham and his boys. They were hanging out on the third floor. I "kindly" let them know of my displeasure of them doing that. Well maybe my tone and wording weren't kindly projected. The particulars of my actual verbiage are neither here nor there. Read that previous sentence in the [7]T.I. intelligent,

[7] T.I. is a famous rapper that sometimes uses a tone and verbiage that's not considered the norm for the stereotypical rapper.

articulate, gangster personality voice. They took off down the stairs after me! I then took off also. Yeah, I took off. It was maybe5 of them. I remember thinking that I couldn't run outside because the car was too far away. They would have killed me in the parking lot! I then remembered where my buddies would hang at over in the South Gym. I began to run that way. It was this long alleyway that led to the Gym. My friends hanging out in the South Gym heard the commotion. I got closer and could see them running my way. Approximately halfway down the hall, the numbers evened up. This scene from the Movie [8] (Juice) comes to mind. 2Pac is surrounded on this fence. He then notices his crew across the street coming to help.

---

[8] 1992 Movie about friends where 2Pac is one of the starring actors.

Chapter 5
Kids of Heroes

Hi, I'm Janet. My Dad was killed in Iraq and I'm sorry for your loss. I know this statement will have you scratching your heads. It's my response whenever people find out about my Dad and say to me, "I'm sorry for your loss". I simply reply to them, "I'm sorry for your loss too".

Over the years I have learned to genuinely mean this whereas younger me would say it as total sarcasm. It has always bothered me that my Dad never made me feel that there was anything special about our bond. This pain continues as I look at old videos of birthday parties where someone has a camera saying to me, "say hi to your Dad sweetie". 22. That's the number of videos of the different events in my life where I was told, "say hi to your Daddy sweetie". As military kids, you learn early in life the importance of the job of your parents.

This importance is emphasized during the times of War. In schools we're taught how heroic our parents are. We hear the word hero so much we begin to associate it with everyday good deeds. In retrospect, I realize how much mental strength is expected from elementary-aged kids of Heroes.

It's painful and confusing, the different energies running through our little bodies. We hear the talk about how heroic our parents are and say to ourselves, "well if you're that much of a Super Hero, why can't you see my pain".

We aren't taught how to relay this message in a manner that leaves both the parent and the child with an understanding of each other's pain without making one side feel like they're the problem. Kids shouldn't have to go through this type of psychological stress without the proper, nor updated programs in place. The thing that complicated the issue for me was the time my Dad came home from a deployment to the Middle East. He told me about all the displaced kids they helped. He showed me videos of himself giving out chocolate and playing soccer/football with those kids. The look in his eyes was the same look I thought was a way in which only I could make him feel. That's the look that made me feel like I was his world.

Seeing how passionate he was about his job and the love he had for it confused the younger me. The pressure of this confusion would only increase as I got older. I struggled with the memories of the life lessons I learned from my dad. I would begin to see them as worthless. I went from sad to enraged. I thought about the lessons my Dad would teach me about people and being truthful. He would often say to me, "when someone shows you who they are, believe them". He taught me to love all people as we are all Spirits' creation. He taught me to stand on my own and not be a follower. To think for myself. I went through a phase of hating my dad. I would see him as a spineless hippocratic coward.

How could he teach me about believing people when he got so much joy following the orders of leaders that constantly lie to him. Especially the lies about reasons behind going to War? How could you kill a child's family and think that giving them chocolate and playing football together will heal their wounds. How can you not expect that child to grow up with hate in their hearts? There was a President that started a whole war because a country tried to kill his dad! How could you enjoy serving knowing that upon completion of service (once used up), soldiers suffering from mental issues will only have more pills stuffed down their throats by organizations that are supposedly in place to help them? They would be forgotten and left in humanity's gutters. Maybe this is their Karma. Why is my dad so brainwashed. I would often think this to myself. To my Dad, I love you. Also to my Dad, F*** YOU. I only wish for you to be here so that I can express those sentiments to you myself! Practice what you PREACH is what I would say to him today. Experiencing such a confusing heartbreak at a young age can remove the kid element from elementary-aged individuals. This forces us to deal with adult situations which we can't complain about.

Often I would say to my Dad that I didn't want him to leave. He would assure me he'd always come back for me. I didn't care about him coming back for me. I WANTED HIM TO BE THERE FOR ME AND GO BACK FOR THEM. I needed his saving. I wanted his unconditional love. My mind is full of memories of the times my Dad and I would be having our father-daughter time. These moments would often be interrupted for him to go off and help one of his soldiers. The other aspect of this confusion is the fact that our parents aren't aware of the fact that there's an issue. They're made to believe that to be good leaders, you must know everything about your subordinates. You must take care of them in all aspects of their lives. Good leaders like my Father took their jobs seriously. A part of me truly believes that he genuinely wanted what's best for everyone. My Father saw himself as a provider and always protected me, as far as he thought. He wasn't aware of the psychological damage his neglect caused me. I guess it was his inner demons which ultimately led to him committing suicide. My Dad's suicide note further complicated my issue. If ever I'm to have a Dad again, I will have to accept the one the thing that's at the root of my suffering. I'll have to expect some other child to share their Dad with me. The one thing I've always wanted was NOT to share my Father. As ironic as it may seem, my dad left a heartwarming message in his suicide note. He assured me his longtime friend Timothy will be more of a father to me than he ever could. The meeting of [9]Lucas's long lost friend Timothy was a blessing. Our chance encounter had saved both our lives. They were more connected than we could have EVER imagined. Timothy's birth name is Will. He changed it after High School to escape his father's shadow. He was tired of always hearing about how big a hero his father was. Will's Father also put service before family. Like myself, Will only wanted his Father to be there more. Wills Father hadn't been aware of the psychological damage the neglect caused Will and himself. Finding out that he had been neglecting his family's battles was too much for Will's Dad to stomach. Will shared his Dads suicide note with me and it read as follows;

---

[9] I would often show my disdain for my dad my calling him by his first name.

Dear Will,

I dont know where to start this letter but I'm going to try. I let my job blindfold me to my true purpose which is protecting you. You are right in all of your complaints as I should have been there for you. Will, dont ever think that any of my shortcomings are of any reflection on you! The past few days have opened my eyes to the realization that as one battle of my life (the completion of service) is coming to an end, another is only beginning. Will, I dont even know where to begin to help you in your suffering so I'm going to help you the only way I know. I remember missing your 16th birthday due to pre-deployment training. It wasnt until recently that I discovered the importance of these special events in your life. I brought joy to hundreds of other kids only to not see my son's hardships. I dont deserve a rekindled relationship where you are to be made to Love me again. You need what I thought had brought me so much relief during the multiple deployments to War. The violent death of those deemed deserving. Now is my time for the neglecting of your number 1 job as a Father! This had to be done as I'm no longer deserving of life in this form! This is what your Mother would have wanted.

Everything they've told us is lies!.

Everything they've told us is lies!

Everything they've told us is lies!

Goodbye Will!

CHAPTER 6
Love :

Love. Love. Love (deep sigh). Before I get into the painful tragic loss of my one true love. The story of how magical our initial meeting was. The story of how losing her stripped me down to my core. This is the story of how she was taken from me so abruptly and my seemingly never-ending battle to gain the mental strength to rejoin "social norms". We planned to do so much together. The Universe gave her to me and war took her away. Before I get into this story, let me help you understand the level of Love I'm speaking of. Take a few seconds and close your eyes. Imagine the

Universe granting your wish of the person of your dreams. Your Twin Flame. That's what she was to me! She was my everything. From our initial meeting, there was a connection. No butterflies. No anxiety. The feeling was that of having known each other before. We both knew it was special. The Universe gave me a jewel. When I first laid eyes on her, she had on this blue velour jumpsuit with NBA patches all over it. The Knicks patch was front and center. I was a die-hard Knicks fan so this was certainly a sign. Long pause...Ya know what! F*** you people and this story. You all don't even like me. You laugh at me thinking I'm homeless. They see me on the street in different spots and sit money by me thinking I'm out there because I need it. I DON'T. When she was killed I couldn't take it! I lost all desire for any type of connection. I would sit at different intersections I thought she may show up and would see me! My former Army

Section Leader would stop by and visit me. He's one of the only people that understood my love for her. He convinced me to look for her in this physical 3D realm rather than follow her into the afterlife. He understands me and my mission to find her. He often drives by and waves as he understands me needing time to be alone. He's an awesome leader and a great friend. I think that subconsciously, I'm sitting at intersections I know he will drive by as I find myself looking for her in Houston and not her hometown of Los Angeles. Every time I have to tell this story, it hurts that much more than it did the previous time! Do you know how this feels? To have something you want so badly end so abruptly. Something that just leaves you hanging! That's what life did to me when it took her! The UNIVERSE gave her to me and WAR took her away. I lost my Angel in a Fallen Angel. This hurts too much. F**K THIS STORY!

About the Author

I was intrigued at an early age about being a soldier. Numerous members of my family served in the Armed Forces. Most noticeably were my Uncles Peter and Bobby Henry Sr.

There are more but these were two I would be around every day. My uncle Peter has a picture of himself in his Army dress uniform. It hangs on the wall in his living room. I would always admire the focus and look of determination in his eyes.

My Uncle Bobby Henry Sr (whom Soul has transitioned home) stories were epic. There would be numerous times we would spend all night at Uncle Peter's house shooting pool. You could hear the loud music from the club a few blocks down. I actually preferred hanging with these two. The stories from those nights were mind-blowing.

I would join the Army on October 24, 2000. Four days before my 20th birthday. I didn't want to spend another birthday in the hood. I desired to get away and see the world! After the completion of high school, I had no desire to attend college. I seen myself as having blown a year of my life doing nothing. The biggie song - Party and BS - would be how I would describe that year after high school.

I attended Basic Training at Fort Jackson in South Carolina. Upon completion of Basic Training, I attended training in Dothan Alabama at Fort Rucker.

After Rucker I headed to Airborne Training at Fort Benning in Columbus Georgia.

After Airborne training, my first duty station would be Fort Bragg in Fayetteville North Carolina.

This was my favorite duty station but would come with some bittersweet memories. I remember feeling guilty as I had friends that were deploying to combat during the times of war in Afghanistan. My Company was chosen to do deployments to the Bahamas. Yeah, the Bahamas. I was a soldier whom my command could trust as the deployment to the Bahamas would be for a small task force. If they only knew how much weed would be smoked during those deployments, they probably would've picked another soldier.

Random Fact that no one knows about myself. Not even the person whom I experienced this with.

I tricked a lot of people into thinking I've been smoking weed since my early teenage years. This isn't true. My first time smoking weed was as a Private in the Army at Fort Bragg. The story about this day is so crazy that you'll think it's made up. It is not. I may incorporate the story into later Volumes. If enough people are curious, I just may give you this story sooner.

After Fort Bragg, I would get orders to go to South Korea. I bought it up to my Commander because I didn't want to go to South Korea. He told me that I could deny my first overseas orders but I would come back up on those same orders. I had to go somewhere overseas. I asked where else could I go. That's how I ended up at Fleigherhorst Kaserne in Hanau Germany in March of 2003.

I would do my first 15-month deployment to Baghdad Iraq in May 2003.

Upon completion of my tour in Germany, I would go to Fort Riley Kansas where I would do a 15-month deployment to Iraq. I would start my separation process from the Army upon return.

I remember thinking that I needed to get out because I was tired of the deployments to the Middle East.

I separated from the Army in May of 2009 only to find myself doing contract work in

Afghanistan less than 8 months later. I would go on to spend more time in the Middle East after separation from the Army than I did while enlisted. The next 7 years would be spent working in the Middle East.

Words of Encouragement

Thank you for reading my book. I mean that wholeheartedly. I would like to share with you all something that resonates so much in my soul. The encouraging words you send through the different social media outlets help. You probably will never meet those individuals like myself whom those words may talk off a ledge or help ease our minds. The biggest issue with those encouraging words is the sentence most of us type before that heartfelt message. That sentence is as follows:

To whomever may need it!

Regardless of how encouraging the message, everything after that statement is rarely felt. It can come off as very condescending. Especially to people that are going through the problems your message speaks directly to. What is heard from the hurting and depressed is, "I do NOT need this message. My life is good. This message is for whoever is f'd up". It's like we're going out of our way to let it be known we don't want to be associated with people with problems. In my experience, It's equivalent to driving by a homeless person and throwing a loaf of bread out the window. We're like, "thanks but damn bro". I've been on both sides of this spectrum. I've been the one reading those words and feeling even more like I was being kicked while I was down. I've also typed that statement prior to sending many encouraging messages. I think it's our subconscious. We aren't trying to kick someone while they're down. Subconsciously, whenever we're doing well in life, we want people to know it. We're a caring people. I think that we aren't aware of how that statement can further some people's anguish. Thanks again. Hopefully this book inspires you all to join me on this journey to bring light to the issues faced by the children of our Heroes!

Acknowledgements

Thank you to my friends from those early years in my military career. My Ft Bragg Days. The Smoke Bomb Hill crew. We were called the Table Ninjas. ➤Mo ➤PDash7 ➤Dave/DJ Enterprize ➤Debo➤Canti. Also my Smoke Bomb Hill family Smitty, Quita, Vicki, Penny, Russel, Ayesha, Mia. Zo, I remember when someone stole your new Honda from in front of the hangar while we were in the field training. I still think you set that up. Charles from Laurinburg, you weren't serving yet, but would make that ride to Bragg and chill with us. Jay Faulks, my buddy. A loyal friend and a good human. There's many more but these are some of my closest. Shawn Stanley, my first boss from my early Army days. Thanks for giving me those speakers for my truck when I was a young broke soldier. You bought me to your house, introduced me to your wife and hooked them up yourself. Your calm persona was contagious and very much needed as I was the total opposite during those days. You kept us reeled in. I remember how you would cover for me at times when I would be working the night shift. The guys would come to get me to go to The Hill and fight.

I was seen as one of the boxing gurus on The Hill. Whenever another fighter came to Bragg, my friends would get me to fight them. Not so much a street brawl. We had gloves and would often drink and smoke afterward. Leesa Briles, you were the best Manager. Thanks you! Quita, you have been such an inspiration. You'll probably never truly know how your energy alone helped me on days I didn't feel like finishing this project. Your story needs its own platform. You are amazing Marquita Mulligan.

Thanks also to my Germany and Ft Riley friends and mentors. Those deployments established a forever bond within us. Vonda

Taylor. You are the epitome of leadership. You would be so tough on us at times but we needed it. You also balanced it out with your loving ways and protection. Feilding, Britt, Wilhite, Tiumalu, to name a few. Carlos Renteria (who Mike who, inside joke with my buddy Renteria), Brandy Hanks. Jeremey Wells (one of the funniest people I know, and I know some characters). Denise, my German friend gute Zeiten. Veronica Staples. You gave me so much inspiration. I remember posting some of my writings on Facebook around 2012ish and you took so much interest in it. It had you intrigued. You were one of the first people to find a genuine interest in my book. Thank you. Talofa Reggie T. Good times. No regrets. Angela Walton aka BG. You have kept me motivated by always inquiring about the different business ventures I've always spoken about. I often think about our bond from those times. You are amazing.

Thanks to some of my former Afghani co-workers and friends. To Sammi, Malik, Johnny, my buddy Faraidoon and the others. Your guys' help in assisting me and keeping our team safe will never be forgotten. Such a hardworking group of guys. My good buddy Nick @nicks_adventures_bolivia, I keep telling my son I'm taking him to your wildlife tours down in Bolivia.

I would like to acknowledge some individuals who probably have no idea of how influential they were to my not giving up on this project and myself. It was during the times of contracting in Iraq when I was being eaten alive by suppressed pain and hurt. The strength that it took to thrive in the infamous Poker Room saved my life! You guys don't know it but the rawness of that room gave me the motivation to keep going.

In life, I was going through the motions. I had given up. NO ONE was exempt from the trash-talking/physical skirmishes of that room. That room wasn't for the weak and we've seen many leave there to say they were never coming back! Junior, Derrick, Doc Drake, Dom, Jamie, Big Tony, Alex, Art

[Angry Mike<<<worst poker face ever] and his crew of ghosts like Nick and → Sneaky Peek >>>>Mr 4-2 off suit - He would call huge pots pre-flop with a 4-2 ←] Female JoJo, Male JoJo (congrats on the girl), Rob Z, Eddie (the guy that introduced me to the table back in 2012ish), Boot, Sue. TT aka Jada aka just kidding, I can't thank you enough.

I want to thank Craig Hoffman. My mentor. Awesome father, great soldier, very knowledgeable leader. If not for you, my career would not have been as smooth sailing as it was. Wonderful person and I promise if you ask any of his previous soldiers, they would say the same. He sacrificed so much for his soldiers and his family. I remember our initial meet. I was this scared kid on his way to airborne school. I didn't know what to expect as I only said yes because of the bonus money they offered me. I began thinking about how could I have decided to go to Jump School when my biggest fear were hikes!

I was sitting in this day-room scared about to be shipped off to Airborne training. I look and see this 6'8 human looking like that Russian Drago dude from Rocky 4. He also had airborne wings on his chest. I walk up and ask him what should I expect from Airborne training. I remember him saying to me, "don't f*** up and you'll be alright soldier". I thought to myself, WTF kind of advice is this! I looked at him with this straight face as he's looking back at me with the same look. I then turn to go back to where I was sitting. As I turned around and began to walk away he called out, "Private Ryan. Surround yourself with smart people". [10]You don't say. This dudes an ass! No wonder he got his butt kicked in Rocky 4! It wasn't until the completion of training when I would think about him again and understand the genius of the advice he had given me. The soldiers that have been through Airborne School know that there's nothing you can say to a young soldier to prepare them for what they're about to experience. You can do everything right, but if the soldier to your front or rear messes up, you are caught between a rock and a hard place. In the sky!

I want to do something a little different. I would like to thank some very beautiful online energies. Some accounts I follow on social media. Most of whom I've never met. Their beautiful energies have picked me up during some tough times while on my journey and in finishing this project. On Instagram the page @law_of_positivism #dogsofinstagram

[10] sarcastic reply to advice that isn't very helpful.

@thefunnyintrovert @iamketurner Ke you're so dope! @angelanissel You're such a dope writer. You're very intelligent and extremely funny. I love reading your posts. @slimthug love the raw motivation brother. Your city-wide contributions speak volumes to your character. From riding bikes around the city, hanging out with fans, riding ATVs in the mud, to showing everyone you meet love. You always have everyone on your projects, doing what you can to keep the togetherness flowing. You are the King Of Houston! That's how you can have a million followers just here in Texas! To my big cousin K Henry. Thank you for your sacrifices. You were an awesome role model for us growing up! Thanks to all you light-filled magical beings! You Servants and Guardians of Light! Namaste!

To my Wonderfull kids. Syanna, Jorja, Marley, Zoey. I love you. We did it bro!

P2 - 2005

<u>P3</u> - <u>Kabul International Airport 2010</u>

Traveling through this airport is one of the wildest experiences I've ever had!

P4-Sunrise in one of my favorite places

P6 - one of my fav morning views

Elliott Ryan

Bitter or Better? Your choice!

Conversation between reenlistment officer and a soldier around July 2007.

Re-up Officer: Staff Sergeant Rions! The Army is offering you a reenlistment bonus of $36,000 for 6yrs. This is an awesome opportunity for you and your family. Your wife is pregnant with your 2nd child and we are scheduled to deploy here in the upcoming months.

Staff Sergeant Rions: I'm getting out Sergeant. I only have a month left on my contract. Thank you anyways.

Re-up Officer: What do you mean you're getting out! We are deploying here shortly and you are stop-loss. You can't get out until after the deployment. You need to think about your family.

Staff Sergeant Rions: I am thinking about my family. I understand the stop-loss process and I'm looking forward to this last deployment. I can save some money before getting out. I want to be around for my kids. I have plans on starting my own business. I love the Army and appreciate my time.

Re-up Officer: You're not being logical. This money will help your family. Your wife is pregnant.

Staff Sergeant Rions: Explain to me how this bonus works.

Re-up Officer: You know how this works Staff Sergeant Rions. You will receive half now and the remaining upon completion of the 6 years. $18,000 now and the remaining afterward.

Staff Sergeant Rions: How much after taxes?

Re-up Officer: Roughly $15,000

Staff Sergeant Rions: That isn't worth the next 12 years of my life making $25000 a year.

Re-up Officer: Get it off your chest Staff Sergeant.

<u>Staff Sergeant Rions</u>: If I reenlist for 6 years, that'll put me at 15 years. It wouldn't make much sense to get out after that. Which would mean at least 4 more deployments and roughly 10 more years physically absent from my family lives.

<u>Re-up Officer</u>: Exactly! Retirement would be right around the corner. The retirement check coming in would allow you to find a simpler job. This would allow you to spend more time home with your family.

<u>Staff Sergeant Rions</u>: What do you mean by another job?

<u>Re-up Officer</u>: When you retire, your retirement check can go to your mortgage which is most people's biggest bill. Therefore you wouldn't have to stress about bills on your next job. You can find something that lets you be home more often with your family.

<u>Staff Sergeant Rions</u>: You're telling me

that after 20 years I wouldn't be able to relax and depend on my retirement check.

Re-up Officer: Of course not. You would have to find another job.

Staff Sergeant Rions: Define retirement! What sense does it make for someone's reenlistment pitch be, when you retire you will have to find another job, work another 20 years after you've already given this job your all and your body is broken down. I love the Army and only want to do my time and get out. The Army has prepared me to be a business owner! I don't understand the confusion. I only want to do something about soldiers having to do 20 years, retire only to have to find another job because they can't take care of their families on that retirement check.

Re-up Officer: I'm referring you to the S3 Sergeant Major. You're not thinking logically!

S3 Sergeant Major: Staff Sergeant Rions! The re-up officer has given me some troubling news. What's your issue?

<u>Staff Sergeant Rions</u>: I don't have an issue. $12,000 isn't worth missing 10 more years of my kid's life. For him to say that I can retire then find another job is an utterly ridiculous reenlistment pitch! As your retirement approaches, I'm certain you can understand this SGM.

<u>S3 Sergeant Major</u>: That's not something you can fix Staff Sergeant! Stay in your lane! What you need to worry about is the here and now and your family. Your wife is pregnant and $36,000 can go a long way.

<u>Staff Sergeant Rions</u>: We all know it's not going to be a full 36 SGM! I can save $12k during this deployment.

<u>S3 Sergeant Major</u>: The additional money can still help your wife.

<u>Staff Sergeant Rions</u>: That's just enough money to get me into more debt SGM. About my wife and her pregnancy. Can I talk for a second about that? I'm scheduled to leave here in August with the advance element. My wife is due in September and I was wondering if I could stay back until the birth of my daughter? I can be on the next plane to Kuwait to meet up with the Unit.

S3 Sergeant Major: Your request is denied! Staff Sergeant!

Staff Sergeant Rions: SGM, the only thing I'm asking is to stay back so I won't miss the birth of my daughter. I know how this works. The Unit is still going to be in Kuwait. We aren't going to push forward yet. I've already spoken with Staff Sergeant Evins and he's more than willing and qualified to take my spot on the advance element. We are still going to be in Kuwait. I can meet you all in Kuwait.

S3 Sergeant Major: Request denied! I will schedule you to be the first to take your leave. You can see your family then.

Staff Sergeant Rions: This isn't my first

deployment. This is your 1st deployment! You're going to make me miss the birth of my daughter, only to see her a month later. Take my leave at the beginning of a 15-month deployment! Come back and miss the next 13 months of my kid's life over there. I've done a 15 deployment. I wouldn't wish for my worst soldier to spend 13 consecutive months deployed! Maybe because you haven't deployed before are you not understanding what you're doing to me. I'm supposed to believe you want me to reenlist because you care for my family. All of this because I don't agree with the modification to the definition of the word RETIREMENT!

Elliott Ryan

THE HATED HEROES

Vol II

Retire so you can get a job!

Elliott Ryan

Responsibilities

(Colonel Brown)

Retirement has come and I sit alone in my rocking chair on the porch in my secluded cabin in the woods of North Carolina. I have never been more confused. Numerous shots of whisky is only adding to the pain from the Love I received from my mentor. Reflection has shown me that the love and fathering I received came at the expense of a kid whom should have been given it instead! I don't know if I hate my mentor or if I hate that I love him for that. Either way, there's a kid who didn't have his father in his life because of me! A father's job is not to abandon his kid. I sit here in total confusion with no one! Whiskey shot after whiskey shot. Shotgun on the side of the rocking chair with a Kimber 1911 in my lap. Everything I've been told has been a lie! Everything I've been told has been a lie! Everything I've been told has been a lie!

🖤

This picture of my oldest daughter Syanna was taken in 2008. I was on deployment to the Middle East. The framed picture was taken during my first deployment to Iraq in 2003. This was her favorite picture. She would sit and stare at it while I was deployed.

I am loving consciousness!
I am loving consciousness!
I am loving consciousness!

Yes, you are!

It's too late! Love has already been sent!

Elliott Ryan

143

Suicide is never the answer!

Suicide is never the answer!

Suicide is never the answer!

Things do get better!

I love you!

Elliott Ryan

MY FIRST TIME

The Fort Bragg weed story

It was around 2001/2002 and a few of the guys and I had become good friends. We were the life of the party on the [11]Hill. Everyone would come to our table because there was always alcohol and music. Some people sharing the area gave our group a nickname.

I would often pull my truck onto the grass, just outside the entryway to the barracks. We used it as the stereo, blasting music throughout The Hill. I had two big speakers and the music just sounded badass from my truck. This was illegal but the Military Police knew us and allowed the show to go on. Periodically, they would make me move my truck. For the most part, they would be cool. They would often pose for pictures with us. During that timeframe, Fort Bragg resembled the atmosphere of a party crazed college town. The only difference was that guns were more accessible here. There were all facets of

---

[11] Smoke Bomb Hill - An area on Fort Bragg

craziness! Bragg was infiltrated with violent gangs, extremists, far left, far right, mentally deranged, addicts, sellers, and every kind of unbalanced individual you thought a life in the military would have distanced you from!

This wasn't our crew. We weren't with malicious intent. We were young and wild. Our fun wasn't rooted in evil. We had our [12]"wrecking crew" ways.

One of my friends was a DJ and had the turntables and studio set up in his room. His not having a roommate made it cool too. Directly across the hall was another friend's room. He was an islander from Jamaica Queens NY. He would often cook this amazing curry for us.

When you enter the first floor foyer, there were 4 rooms. Two rooms on the right and two rooms on the left. There was a washer and dryer directly to your front. The room doors

12 Phrase taken from Tupac in the 1992 movie Juice.

would be open as we talked in the foyer and outside. Only a select few were allowed in the foyer! Everyone was invited outside. We were talking in the foyer and someone bought up weed. My friend Stash knew where the spots were. My truck was being used for the music so we would take his car. I thought we would go there, get the weed, and come right back. After a short trip to the [13]Merc, we ended up at one of Stash's weed man's house. I was surprised because it was a nice quiet neighborhood.

We pulled up and were greeted by this cool Rastafarian. He had, as they describe their hair, a forest pon me head. As we sit there, Stash and the Rasta begin having a conversation. I've always came off as quiet. You get to know me and realize that couldn't be further from the truth. The Rasta didn't really

---

[13] Of one the rougher parts of town in Fayetteville North Carolina

communicate with me. He would look "side-eyed" at me a few times. I could sense he wasn't sold on me just yet. After a few minutes of conversation, the Rasta goes into the back and brings out this red 5-gallon bucket full of weed. He sparked up a joint and began smoking. He would then pass the joint to Stash. Stash would smoke and then came the moment of truth.

Ever since the Rasta sparked the joint, I was freaking out internally. I've never smoked weed before. I didn't know what I was going to do when it was my time.

(*This funny story came to mind about the time we were younger and one of my brothers smoked weed on a Sunday morning before Church. He started freaking out in the middle of church service while playing the drums (a hilarious story itself)!*)

Stash would go to hand me the joint and I replied, "I'm good". The Rasta almost immediately began to go off. His accent would thicken as he would be screaming at Stash, while pointing at me. I couldn't understand completely what he was saying. I caught a few words and realized that I really needed to hit this joint. Stash thought it was funny until the Rasta goes into the back room and comes out with an AK-47.

Those familiar with an AK-47 know these guns have hair triggers. The last weapon you want an unstable, irate individual to point at you is an AK-47. The slightest brush against the trigger will discharge the weapon. The Rasta came out pointing the rifle at Stash. He would be going back and forth, pointing the gun at Stash and myself. I was like "WHOA WHOA DON'T SHOOT US. PLEASE"!

I would be using my hands as shields, going back and forth blocking my head and legs. As if my hands were going to block the bullets. The altercation resembled the Ice Cube movie,

(14All about the Benjamins). The scene when Mike Epps finally found his wallet on the boat. The man sneaks up behind Mike with the gun. Hilarious Movie!

I then told the Rasta to hand me the joint. He began to calm down a little. He grabbed the joint, re-lit it and handed it to me. I remember thinking, this shit is the size of a freaking pickle. I'm about to die.

I did smoke black and mild cigars at the time. I figured if I would smoke it like that, all would be good. I began to smoke the joint. I just hit it over and over while inhaling and being fancy. Sucking the smoke through my nose and blowing circles. I did this for like two minutes straight. I remember the Rasta calming down and saying, "with lungs like that, this can't be your first time smoking weed". Stash would have this silly dramatic look on his

---

[14] 2002 Ice Cube movie starring Mike Epps and Eva Mendes

face, nonverbally communicating the message, “Kleazy, what the entire f*** just happened”? We would finish the joint. I remember sitting there listening to this conversation (not involving me) going on in my head. Myself was saying to my other self, "something is happening to you but you must stay cool dude. Don’t start tripping like your brother”.

The Rasta’s hair started to spell words. This threw me for a loop. I remember thinking, “Yup you’re gonna die. Big dummy”! After a few more minutes, the Rasta had completely calmed down. We would then get our weed and begin to head out. I remember Stash laughing the entire time.

Once in the car, Stash looked at me with this hilarious grin and asked if I was alright. I replied, “yeah”, and he then pulled off. After a couple of blocks, the car began lifting off the road. This caused me to sit up abruptly. I placed both my hands on the dashboard of the car and stared out the windshield. Stash could tell something was off, and asked again, "Kleaz, you good man"? I replied, "I think I need to go to the hospital". Stash would burst into laughter. I was not playing!

I was able to gather myself until we made it back to the barracks. There was a party going on and my car was still on The Hill. I knew I was in no shape to be around other humans. I couldn't feel my feet. It didn't feel as if I were walking on solid ground. I had Stash to drop me off at my room. My room was the length of a football field from where the others stayed. I could stand on my 2nd-floor balcony, squint my eyes, and see if anyone was at the table on the Hill. My roommate was a Devil worshiper. When I initially got to Bragg, I walk in and see this dude. He was a tall, long haired, unclean, pale skin human. We had a "discussion" when I arrived. He thought his shenanigans would freak me out. I remember telling him that as long as he kept all his skulls, bones, and other artifacts on his side of the room, there would be no problems. I don't care whom you pray to. The rooms were maybe a couple hundred square feet. As you enter the room, there's a narrow walkway. The cooking area is to your immediate right. The bathroom is straight ahead past the kitchen area. As you take a couple paces forward, the roommates sleeping area is to your left. You take another couple steps and I was in my sleeping area. The areas were separated by our two tall entertainment centers. His facing his area and vice versa. We would have shower curtains hanging from the wall to the entertainment center to give ourselves some privacy. I get back to my room and my roommate wasn't there. He would always leave his curtain open so people can see his worshiping artifacts. I get to his area and pause while examining his artifacts. After a brief moment, I then proceeded to my side. As

I get past his side I hear, "you're gonna die dumb ass". I stop, rewind my steps, lean back with only my head peaking into his area to see if maybe I overlooked my roommate being there. There was nothing. I proceed to my side, lay on my bed and I hear it again. I was like, WTF! I then burst into laughter because it was the funniest thing ever. I lay on my bed with my eyes open because every time I close them, I would spin backward. I was extremely high. I would lay there laughing in a full on roasting session, going back and forth with these hilarious skulls. They laughed at me repeatedly saying I was a dumb ass and was going to die.

Rainbows Are Not Sexual, LLC

RANS

Guardians Of The Rainbow

Elliott Ryan

Through it all, I remained hungry. Special Thanks to Dr. Les Brown. You are appreciated!

Elliott Ryan

Made in the USA
Monee, IL
05 November 2026

67378288R00093